BAKER'S LIST

OF "COLORED" INVENTORS

BAKER'S LIST
OF "COLORED" INVENTORS

Over 400 Inventions by African Americans (1821–1980)

Reggie K. Smith

Copyright © 2025 by Reggie K. Smith

All Rights Reserved.

Design and production by Reggie K. Smith.

To all the "Colored," Black, Negro, Mulatto, African American, slave, and free inventors who, for whatever reasons—legal, societal, cultural, and financial—were prevented from receiving patents for their inventions, therefore impeding their economic gain and recognition.

TABLE OF CONTENTS

ACKNOWLEDGEMENTS

I wish to extend my deepest gratitude to Henry E. Baker (1857–1928), whose pioneering work in chronicling African American inventors has inspired this publication. The Colored Inventor: A Record of Fifty Years stands as a testament to the ingenuity, perseverance, and brilliance of Black innovators. His dedication to preserving history has made it possible for future generations to recognize and celebrate these invaluable contributions. Without his meticulous research and commitment, this collection would not be possible.

You may find free published Kindle Edition reprints of Baker's book here:

https://www.HEBakerslist.com#BakerBooks

INTRODUCTION

Henry E. Baker (c. 1902)

"It is not so apparent, however, to the general public that along the line of inventions also the colored race has made surprising and substantial progress; and that it has followed, even if 'after off,' the footsteps of the more favored race. And it is highly important, therefore, that we should make note of what the race has achieved along this line to the end that proper credit may be accorded as having made some contribution to our national progress."

—Henry E. Baker, 1913

At 3:00 a.m. on December 15, 1836, a fire broke out in the Blodgett Hotel in Washington, D.C., then occupied by the United States Patent Office. It would be the first of two significant fires in the history of the Patent Office. The fire destroyed patent documents as well as physical models of patents submitted by inventors to the Patent Office for approval.

Afterwards, the Patent Office, together with the U.S. Congress, initiated policies that changed the way it handled record keeping. The Patent Office began assigning numbers to patents and required multiple copies of supporting documentation. They categorized any known patents prior to the fire as "X-Patents" by assigning an "X" with a number.

A second fire occurred in 1877. The Patent Office, again, suffered losses of records and models and subsequently made policy changes—one being the dispensing of requiring models to be submitted with patent applications. It was also the year Henry E. Baker began employment at the U.S. Patent Office. Henry Edwin Baker of Columbus, Mississippi, is the reason we've come to understand the contributions of African American inventors in the 19th century and early 1900s. He started at the U.S. Patent Office as a copyist. During his employment, he continued his education, graduating from Ben-Hyde Benton School of Technology in Washington, D.C., and soon, Howard School of Law. In 1902 he attained the rank and title of Assistant Patent Examiner. In addition, he earned the trust and respect of the commissioners, allowing him to oversee one of the most significant projects in the history of the U.S. Patent Office and in the achievements of African Americans in the field of invention.

It has never been a policy of the U.S. Patent Office to record ethnic or cultural identity on patent applications. One exception Baker noted was an entry made in the patent application of inventor Henry Blair, listing him as a "colored man." Henry Blair was the second African American to receive a patent for his seed planter in 1836. The patent examiner likely noted this, given the legal rulings governing patents issued to slaves and freedmen across slave-holding and free states. The first African American to receive a patent was tailor Thomas Jennings for "dry-scouring," otherwise known as dry cleaning.

On three notable occasions, however, the U.S. Patent Office agreed to provide whatever information it could regarding patents issued to "colored" or "Negro" inventors to those requesting such. Three organizations of note that made such requests were the United States Congress via the efforts of U.S. Representative George Washington Murray, then the only Black member of Congress in 1894; the Paris 1900 Exposition; and the Pennsylvania Emancipation Exposition of 1913. With the backing of the U.S. Patent Office, Baker helped organize a campaign that included sending written correspondence throughout the country to patent attorneys, newspapers, corporations, factories, community leaders, and others, requesting feedback regarding known African American inventors. His efforts resulted in a list of over 1200 submissions, of which Baker verified and noted 800. Baker distributed available lists of inventors and inventions to the parties mentioned above for their purposes.

U.S. Representative George Washington Murray of South Carolina was an African American politician, inventor, and educator. He received multiple patents for agricultural tools and was elected to the U.S. House of Representatives in the 1890s. During his tenure, he fought against disenfranchisement laws that sought to suppress Black political participation. It was Representative Murray who acquired a list of African American patent holders from Henry E. Baker and boldly read the list into the Congressional Record so that history may bear record of the accomplishments of African American inventors.

The Paris 1900 Exposition (Exposition Universelle) was a grand world's fair held in Paris from April 14 to November 12, 1900. It celebrated the achievements of the 19th century while showcasing innovations that would shape the future. The event attracted over 50 million visitors and featured groundbreaking technologies like escalators, electric cars, and talking films. Iconic structures such as the Grand Palais and the Petit Palais were built for the exposition, and it also brought international attention to the Art Nouveau style.

Notable among the attendees was American sociologist, historian, and civil rights activist W.E.B. Du Bois, who curated the American Negro exhibit displayed at the Paris 1900 Exposition. Henry E. Baker supplied Du Bois and his committee with a list of African American inventors and inventions for the exhibit. Paris Exposition judges awarded Du Bois a gold medal for his role as "collaborator" and "compiler" of materials for the exhibit.

South Philadelphia hosted the Pennsylvania Emancipation Exposition of 1913 to commemorate the 50th anniversary of the Emancipation Proclamation. The exposition featured exhibits, parades, athletic contests, and musical performances, celebrating African American progress in education and industry. Despite racial tensions and criticism from some white-owned newspapers, the exposition attracted around 100,000. The event showcased resilience and achievements, leaving a lasting impact on the community. Baker contributed his latest findings on African American inventors to the organizing committee of this exposition.

Henry E. Baker was a visionary who overcame racial barriers through perseverance and the transformative power of knowledge, becoming a tireless advocate for African American inventors. His legacy is one of ingenuity, empowerment, and equality, as he worked to document and celebrate the contributions of the underrepresented, fostering recognition and justice through his groundbreaking efforts.

Baker was able to confirm a significant portion of the several thousand responses, despite various obstacles that made it challenging to get patent information from some African American patent holders. All replies were directed to Baker, who went about the task to "trace a clew or clinch a fact" (p. 28).

> "Sometimes it has been difficult to get this information by correspondence even from colored inventors themselves. Many of them refuse to acknowledge that their inventions are in any way identified with the colored race, on the ground, presumably, that the publication of that fact might adversely affect the commercial value of their invention; and in view of the prevailing sentiment in many sections of our country, it cannot be denied that much reason lies at the bottom of such conclusion."
>
> —Henry E. Baker, 1913

The U.S. Patent Office played a huge supporting role in providing access to internal and external resources.

The list herein represents a portion of Baker's work. The Moorland Springarn Library of Howard University hosts the fuller body of Baker's ultimate four volumes. Likewise, the internet is host to a few extract lists—although varying in quality. It is my endeavor with this publication to provide a comprehensive list that preserves the legacies of the featured inventors and ensures their contributions are not lost to time. Lists such as this inspire future generations, particularly young people, to explore STEM fields and recognize their potential for innovation, thereby connecting history with future advancements. Likewise, I provide educators and researchers with a valuable tool that fosters meaningful discussions and encourages exploration of African American history. The list includes the Paris 1900 Exposition submission by Henry E. Baker.

I will reserve a future title for a deeper dive into my fellow Mississippian, Henry Edwin Baker's inspiring life story and his groundbreaking journey from Columbus, Mississippi, to Washington, D.C. The title will showcase beautiful illustrations of selected patents from Baker's list to highlight the remarkable and innovative contributions of the prolific "colored" inventors that Baker passionately spoke of—Baker's list of inventions or an update to be included, of course.

THE BASICS ABOUT PATENTS

Holding a patent historically validates your innovation. It gives you exclusive rights to your invention, meaning no one else can make, use, or sell it without your permission for a set period—typically 20 years for utility patents. Patents provide numerous business opportunities, licensing deals, and potential profits. Being a patent holder establishes your name in the world of innovation, proving that you've contributed something valuable to society. As well, patents incentivize inventors to bring fresh solutions into the world.

To obtain a patent, an inventor must file an application with the United States Patent and Trademark Office (USPTO), ensuring their idea meets the criteria of being novel, non-obvious, and useful—a criterion that has been refined over the years. Early American colonies granted inventors commercial rights to their inventions based on successful appeals to the colonial government. More standardized procedures for patent applications emerged in the late 18th century. These standards typically followed English practices. On a state level, for example, the first general patent act of South Carolina described a patent as follows:

> "The Inventors of useful machines shall have a like exclusive privilege of making or vending their machines for the like term of 14 years, under the same privileges and restrictions hereby granted to, and imposed on, the authors of books."

The first federal patent act of the United States was the Patent Act of 1790. The criteria for passing the examination process were "not before known or used" and "sufficiently useful and important." Updates made via the Act of 1793 yielded the description we use today.

> "any new and useful art, machine, manufacture or composition of matter and any new and useful improvement on any art, machine, manufacture or composition of matter."

Henry E. Baker described the inventive process as "...the endeavor both to create new things and to effect such new combinations of old things as will adapt them to new uses (pg 2)." No matter how genuine patents seem or how well-known they have become, the majority of the patents listed here and around the world are "combinations of old things." Not many are unique. However, drawing from the past is beneficial both here and in life. Whether new or improved, the inventor will state so in the patent application.

Although a U.S. patent is a comprehensive legal document, the standard application basically consists of a specification, claims, drawings, and a cover page. The U.S. Patent and Trademark Office offers a wealth of patent information on their website: https://www.uspto.gov/patents. Likewise, there is no shortage of sites dedicated to educating viewers on the patent process.

If you have ideas and wish to pursue the patent process, here are a few basic steps to get started.

1. **Patentability**: Before you begin, check whether your invention is novel, useful, and non-obvious. Conduct preliminary research to see if similar patents already exist.

2. **The Patent Search**: You may use available patent databases resources such as the U.S. Patent and Trademark Office or international patent offices to help you determine whether your invention is unique and not already patented.

3. **Patent Types**: Choose the type of patent that best suits your invention: Utility (new and useful processes, machines, or compositions); Design (unique appearance); and Plant (newly invented or discovered plant varieties).

4. **Prepare Your Application**: This includes a detailed description of your invention, drawings (if applicable), and claims that define the scope of protection you seek.

5. **File with the USPTO**: You may file a Provisional or Non-provisional patent application. One offers temporary protection while you refine your invention and the other is the full application process.

6. **Examination Process**: The patent office review of your work.

7. **Patent Grant and Maintenance**: If approved, you receive a patent, typically lasting 20 years for a utility patent (or 15 years for a design patent).

Feel free to consult a patent attorney. Costs may vary but expect to pay fees for filing, search and examination, issuing and maintenance, attorney, drawings, and any international fees.

Here is an example of a typical patent from Baker's list.

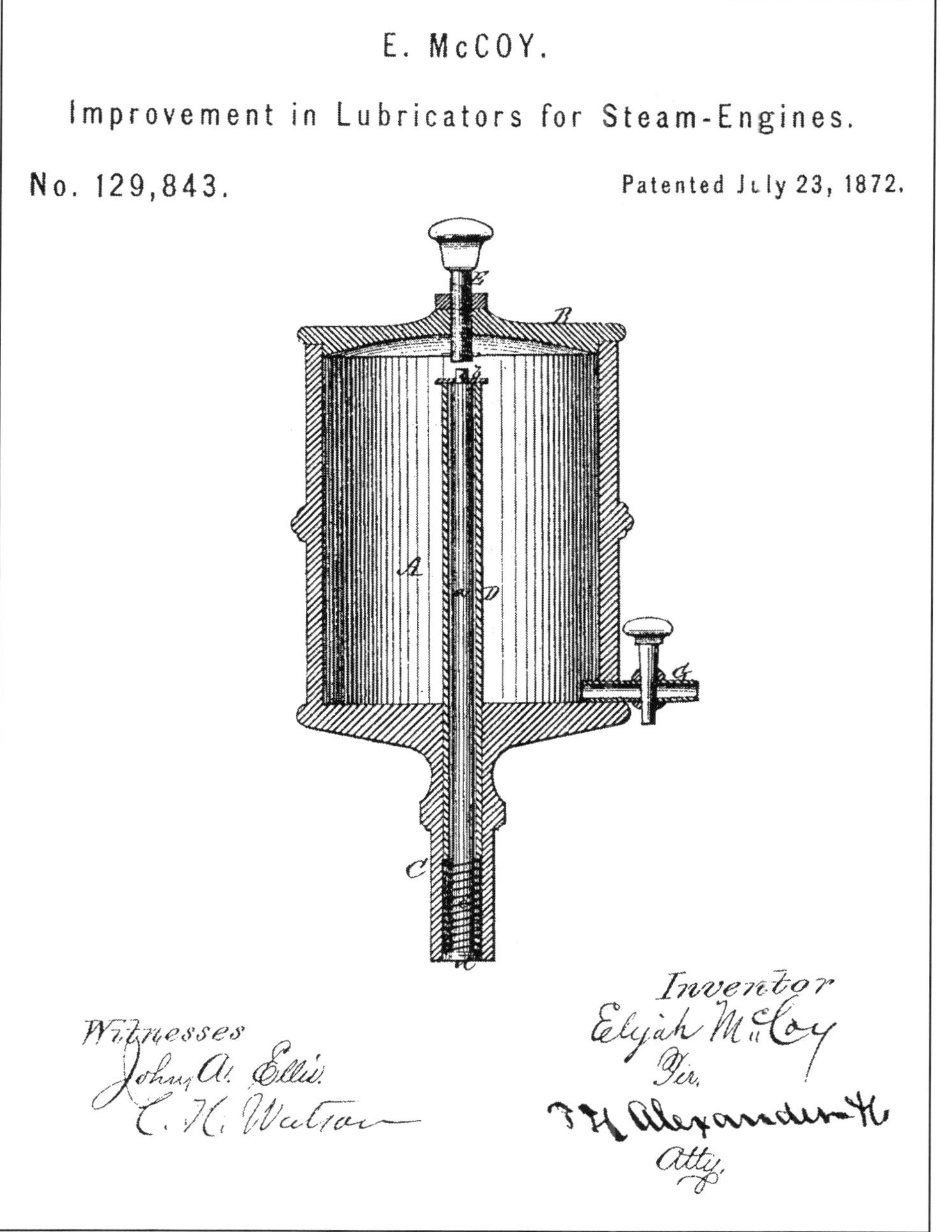

Elijah McCoy patent # 129,843, 1872, U.S. Patent and Trademark Office

UNITED STATES PATENT OFFICE.

ELIJAH McCOY, OF YPSILANTI, MICHIGAN, ASSIGNOR TO HIMSELF AND S. C. HAMLIN, OF SAME PLACE.

IMPROVEMENT IN LUBRICATORS FOR STEAM-ENGINES.

Specification forming part of Letters Patent No. **120,843**, dated July 23, 1872.

SPECIFICATION.

To all whom it may concern:

Be it known that I, ELIJAH McCOY, of the city of Ypsilanti, in the county of Washtenaw and State of Michigan, have invented certain new and useful Improvements in Lubricators; and I do hereby declare that the following is a full, clear, and exact description thereof, reference being had to the accompanying drawing and to the letters of reference marked thereon, which form a part of this specification.

The nature of my invention consists in the construction and arrangement of a lubricator for steam-cylinders, as will be hereinafter more fully set forth.

In order to enable others skilled in the art to which my invention appertains to make and use the same, I will now proceed to describe its construction and operation, referring to the annexed drawing, in which is represented a longitudinal section.

A represents the oil-cup, provided with the cover B. In the center of the bottom of the cup A is a downward-projecting stem, C, to be screwed into the place where the lubricator is to be used. This stem is hollow, and from the same extends a tube, D, through the center of the cup. Within this tube is a rod, *a*, having a valve, *b*, at its upper end above the tube D to close the same, and at the lower end is a piston or disk, *d*, within the stem C. Around the lower end of the rod *a*, between the piston *d* and a shoulder in the stem, is placed a spiral spring, *e*, which forces the rod down, so that the valve *b* will close the upper end of the tube D and prevent the passage of the oil.

When the steam presses upon the piston the valve rises and allows the oil or other lubricating material used to pass out.

In the cover B is a thumb-screw, E, directly above the valve *b*, by means of which the flow of oil may be readily regulated. At the bottom of the oil-cup is a faucet, G, for the purpose of drawing off the condensed steam when necessary.

Having thus fully described my invention, what I claim as new, and desire to secure by Letters Patent, is—

1. The tube D, rod *a*, and spring *e*, in combination with the valve *b* and thumb-screw E and top B, the several parts being arranged to operate substantially as and for the purpose specified.

2. The stem C, tube D, rod *a*, and piston *d*, in combination with the spring *e*, when the spring is arranged in the stem and between the piston and end of the tube, substantially as and for the purpose set forth.

In testimony that I claim the foregoing as my own I hereby affix my signature in presence of two witnesses.

ELIJAH McCOY.

Witnesses:
S. M. CUTCHEON,
W. R. SAMSON.

Granted, this patent is via the U.S. Patent Office procedures in place in 1872; the legal grounds are covered, providing protection and acknowledgement for the inventor. Note the identification of "Improvements" in the Specification section. Prolific inventors like Elijah McCoy would frequently invent improvements to their original or already improved devices.

Obviously, there are steps to take prior to filing a patent application, including the inventive process itself. All inventions start with an idea to solve a problem. If you're interested in some creative fun, explore the TRIZ method of problem solving: https://www.triz40.com/TRIZ_GB.php.

BAKER'S LIST of "Colored" Inventors
Over 400 Inventions by African Americans
(1821–1980)

The following list has been compiled based on the efforts of Henry E. Baker, with support from the United Patent Office, to identify and document known inventions by African American inventors. It contains records that were read into the U.S. Congressional Record during the 1890s and lists submitted to the 1900 Paris Exposition and the Pennsylvania Emancipation Exposition of 1913.

INVENTOR	INVENTION	PATENT	DATE	ST
Abrams, William B.	Hame Attachment	450550	1891-04-14	NY
Albert, Albert P.	Cotton Picking Apparatus	851475	1907-04-23	LA
Alexander, Nathaniel	Chair	997108	1911-07-04	VA
Alexander, Ralph W.	Corn Planter Check Rower	256610	1882-04-18	IL
Allen, Charles W.	Self-Leveling Table	613436	1898-11-01	PA
Allen, James B.	Clothes Line Support	551105	1895-12-10	NJ
Ashbourne, Alexander P.	Process Preparing Coconut For Domestic Use	163962	1875-06-01	CA
Ashbourne, Alexander P.	Biscuit Cutter	170460	1875-11-30	CA
Ashbourne, Alexander P.	Process of Treating Coconut	194287	1877-08-21	CA
Ashbourne, Alexander P.	Refining Coconut Oil	230518	1880-07-27	CA
Bailey, Leonard C.	Combined Truss and Bandage	285545	1883-09-25	DC
Bailey, Leonard C.	Folding Bed	629286	1899-07-18	DC
Bailiff, Charles O.	Shampoo Headrest	612008	1898-10-11	MI
Bailis, William M.	Ladder Scaffold-Support	218154	1879-08-05	NJ
Ballow, William J.	Combined Hat rack and Table	601422	1898-04-29	TN
Barnes, Ned E.	Sand Band	792109	1905-06-13	TX
Barnes, Ned E.	Rail Tie and Brace	815059	1906-03-13	TX
Barnes, Ned E.	Hot Box Cooler and Oiler	899939	1908-09-29	TX
Barnes, Ned E.	Indicator or Bulletin	969592	1910-09-06	TX
Beard, Andrew J.	Rotary Engine	478271	1892-07-05	AL
Beard, Andrew J.	Car Coupler	594059	1897-11-23	AL
Becket, George E.	Letter Box	483525	1892-10-04	RI
Bell, Landrow	Smoke Stacks for Locomotives	115153	1871-05-23	DC
Bell, Landrow	Dough Kneader	133823	1872-12-10	DC
Benjamin, Lyde W.	Broom Moisteners and Bridles	497747	1893-05-16	MA
Benjamin, Miriam E.	Gong and Signal Chairs for Hotels	386289	1888-07-17	DC
Benton, James W.	Lever Derrick	658939	1900-10-02	KY
Binga, M. William	Street Sprinkling Apparatus	217843	1879-07-22	OH

Blackburn, Albert B.	Railway Signal	376362	1888-01-01	OH
Blackburn, Albert B.	Spring Seat for Chairs	380420	1888-04-03	OH
Blackburn, Albert B.	Cash Carrier	391577	1888-10-23	OH
Blair, Henry	Corn-planting machine	0	1834-10-14	MD
Blair, Henry	Corn-harvesting machine	0	1836-08-31	MD
Blue, Lockrum	Hand Corn Shelling Device	298937	1884-05-20	DC
Boone, Sara	Ironing Board	473653	1892-04-26	CT
Bowman, Henry A.	Making Flags	469395	1892-02-23	MA
Bradberry, Henrietta	The Bed Rack	2320027	1943-05-25	IL
Brooks, Charles B.	Punch	507672	1893-10-31	NJ
Brooks, Charles B.	Street Sweepers	556711	1896-03-17	NJ
Brooks, Charles B.	Street Sweepers	560154	1896-05-12	NJ
Brooks, Hallstead/Page	Street Sweepers	558719	1896-04-21	NJ
Brown, Henry	Desk-top File Box	352036	1886-11-02	DC
Brown, Lincoln F.	Bridle Bit	484994	1892-10-25	OH
Brown, M. V. B.	Home Security System Utilizing Television Surveillance	3482037	1969-12-02	NY
Brown, Oscar E.	Horseshoe	481271	1892-08-23	NY
Brown, Oscar E.	Horseshoe	524055	1894-08-07	NY
Brown, Oscar E. & Latimer, Lewis H.	Water Closets for Railway Cars	147363	1874-02-10	MA
Browne, Hugh M.	Sewer or Other Trap	426429	1890-03-29	DC
Browne, Hugh M.	Damper Regulater	886183	1908-04-28	PA
Burkins, Eugene	Breech Loading Cannon	649433	1900-05-15	IL
Burr, John Albert	Lawn Mower	624749	1899-05-09	MA
Burr, William F.	Switching Device for Railways	636197	1899-10-31	MA
Butler, Richard A.	Train Alarm	584540	1897-06-15	RI
Butts, John W.	Luggage Carrier	634611	1899-10-10	MA
Byrd, Turner J.	Improvement in Holders for Reins for Horses	123328	1872-02-06	MI
Byrd, Turner J.	Apparatus Detaching Horses from Carriages	124790	1872-03-19	MI
Byrd, Turner J.	Improvement in Neck Yokes for Wagons	126181	1872-04-30	MI
Byrd, Turner J.	Improvement in Car	157370	1874-12-01	MI

	Couplings			
Campbell, Peter R.	Screw Press	213871	1879-04-01	MS
Campbell, Robert L.	Valve Gear for Steam Engines	728364	1903-05-19	AL
Campbell, William S.	Self-Setting Animal Trap	246369	1881-08-30	NE
Cargill, Benjamin F.	Invalid Cot	629658	1899-07-25	PA
Carrington, Thomas A.	Range	180323	1876-07-25	MD
Carter, W. C.	Umbrella Stand	323397	1885-08-04	OH
Carver, George Washington	Cosmetics and Producing the Same	1522176	1925-01-06	MO
Carver, George Washington	Paint and Stain and Producing the Same	1541478	1925-06-09	MO
Carver, George Washington	Producing Paint and Stains	1632365	1927-06-14	MO
Cassell, Oscar R.	Bedstead Extension	990107	1911-04-18	NY
Cassell, Oscar R.	Flying Machine	1024766	1912-04-30	NY
Cassell, Oscar R.	Angle Indicator	1038291	1912-09-10	NY
Cassell, Oscar R.	Bedstead Extension	1105487	1914-07-28	NY
Certain, Jerry M.	Parcel Carrier for Bicycles	639708	1899-12-26	FL
Chapman, Coit T.	Cotton Planter and Fertilizer Distributer	423311	1890-03-11	SC
Cherry, Matthew A.	Velocipede	382351	1888-05-08	DC
Cherry, Matthew A.	Street Car Fender	531908	1895-01-01	DC
Christmas, Charles T.	Hand Power Attachment for Sewing Machines	226492	1880-04-13	MS
Christmas, Charles T.	Bailing Press	228036	1880-05-25	MS
Christmas, Charles T.	Bale Band Tightener	231273	1880-08-17	MS
Church, Titus S.	Carpet Beating Machine	302237	1884-07-22	MA
Clare, Obadiah B.	Trestle	390753	1888-10-09	IA
Clark, Erastus J.	Nut Lock	308876	1884-12-09	IL
Coates, Robert	Overboot for Horses	473295	1892-04-19	DC
Cook, George	Automatic Fishing Device	625829	1899-05-30	KY
Coolidge, Joseph S.	Harness Attachment	392908	1888-11-13	DC
Cooper, A. R.	Shoemaker's Jack	631519	1899-08-22	OH
Cooper, Jonas	Shutter and Fastening	276563	1883-05-01	DC
Cooper, Jonas	Elevator Device	536605	1895-04-02	DC
Cooper, Jonas	Elevator Device	590257	1897-09-21	DC

Cornwell, Phillip W.	Draft Regulator	390284	1888-10-02	MA
Cornwell, Phillip W.	Draft Regulator	491082	1893-02-07	MA
Cosgrove, William F.	Automatic Stop Plug for Gas Oil Pipes	313993	1885-03-17	NJ
Cralle, Alfred L.	Ice-Cream Mold and Disher	576395	1897-02-02	PA
Creamer, Henry	Steam Fed Water Trap	313854	1885-03-17	NY
Creamer, Henry	Steam Fed Water Trap	358964	1887-03-08	NY
Creamer, Henry	Steam Trap	376586	1888-01-17	NY
Creamer, Henry	Steam Trap Feeder	394463	1888-12-11	NY
Creamer, Henry	Steam Trap	404174	1889-05-28	NY
Creamer, Henry	Steam Trap	457983	1891-08-18	NY
Creamer, Henry	Steam Trap	509202	1893-11-21	NY
Dammond, William H.	Safety System for Operating Railroads	823513	1906-06-19	MI
Darkins, John T.	Ventilation Aid	534322	1895-02-19	PA
Davidson, Shelby J.	Paper Rewind Mechanism for Adding Machines	884721	1908-04-14	KY
Davis, Israel D.	Tonic	351829	1886-11-02	CA
Davis, William D.	Riding Saddles	568939	1896-10-06	MT
Davis, William R., Jr.	Library Table	208378	1878-09-24	NY
Davis, William R., Jr.	Game Table	362611	1887-05-10	NY
Deitz, William A.	Shoe Upper	64205	1867-04-30	NY
Dickinson, Joseph H.	Reed Organ	624192	1899-05-02	MI
Dickinson, Joseph H.	Adjustable Tracker for Pheumatic Playing Attachments	915942	1909-03-23	NJ
Dickinson, Joseph H.	Volume Controlling Means for Mechanical Musical Instruments	926178	1909-06-29	NJ
Dickinson, Joseph H.	Player Piano	1028996	1912-06-11	NJ
Dickinson, Joseph H.	Arm for Recording Machine	1262411	1918-01-08	NJ
Dixon, James	Car Coupling	471843	1892-03-29	OH
Dorcas, Lewis B. and Ryan, Tazwell W.	Stove	868417	1907-10-15	WV
Dorsey, Osbourn	Door-Holding Device	210764	1878-12-10	DC
Dorticus, Clatonia Joaquin	Applicator for Coloring Liquids on Shoe Soles and Heels	535820	1895-03-19	NJ

Dorticus, Clatonia Joaquin Dorticus	Machine for Embossing Photo	537442	1895-04-16	NJ
Dorticus, Clatonia Joaquin Dorticus	Photographic Print Wash	537968	1895-04-23	NJ
Dorticus, Clatonia Joaquin Dorticus	Hose Leak Stop	629315	1899-07-18	NJ
Downing, Gertrude E.	Corner Cleaner Attachment	3715772	1973-02-13	DC
Downing, Phillip B.	Electric Switch for Railroad	430118	1890-06-17	MA
Downing, Phillip B.	Street Letter Box	462092	1891-10-27	MA
Downing, Phillip B.	Letter Box	462093	1891-10-27	MA
Downing, Phillip B.	Street Letter Box	462096	1891-10-27	MA
Doyle, James	Serving Apparatus for Dining Rooms	659057	1900-10-02	PA
Doyle, James	Automatic Serving System	1019137	1912-03-05	PA
Doyle, James	Server for Automatic Serving Systems	1098788	1914-06-12	PA
Drew, Charles R. MD	Apparatus for Preserving Blood	2301710	1942-11-10	DC
Dunnington, James H.	Horse Detachers	578979	1897-03-16	PA
Edmonds, Thomas H.	Separating Screens	586724	1897-07-20	DC
Elkins, Thomas	Dining, Ironing Table and Quilting Frame Combined	100020	1870-02-22	NY
Elkins, Thomas	Chamber Commode	122518	1872-01-09	NY
Elkins, Thomas	Refrigerating Apparatus	221222	1879-11-04	NY
Evans, John H.	Convertible Settees	591095	1897-10-05	OH
Faulkner, Henry	Ventilated Shoe	426495	1890-04-29	MA
Ferrell, Frank J.	Steam Trap	420993	1890-02-11	NY
Ferrell, Frank J.	Apparatus for Melting Snow	428670	1890-05-27	NY
Ferrell, Frank J.	Valve	428671	1890-05-27	NY
Ferrell, Frank J.	Valve	450451	1891-03-14	NY
Ferrell, Frank J.	Valve	462762	1891-11-10	NY
Ferrell, Frank J.	Valve	467796	1892-01-26	NY
Ferrell, Frank J.	Valve	468242	1892-02-02	NY
Ferrell, Frank J.	Valve	468334	1892-02-09	NY
Ferrell, Frank J.	Valve	490227	1893-01-17	NY
Ferrell, Frank J.	Valve	501497	1893-07-18	NY
Fisher, David A., Jr.	Joiner's Clamp	162281	1875-04-20	DC

Fisher, David A., Jr.	Furniture Castor	174794	1876-03-14	DC
Flemmings, R. F., Jr.	Guitar	338727	1886-03-03	MA
Goode, Sarah E.	Cabinet Bed	322177	1885-07-14	IL
Gourdine, Meredith C.	Electrogasdynamic method and apparatus	3449667	1969-06-10	NJ
Grant, George F.	Curtain Rod Support	565075	1896-08-04	MA
Grant, George F.	Golf Tee	638920	1899-12-12	MA
Gray, Robert H.	Bailing Press	525203	1894-08-28	KY
Gray, Robert H.	Cistern Cleaners	537151	1895-04-09	KY
Gregory, James	Motor	361937	1887-04-26	SC
Grenon, Henry	Razor Stropping Device	554867	1896-02-18	CT
Griffen, Thomas W.	Pool Table Attachment	626902	1899-06-13	CT
Gunn, Selim W.	Boot or Shoe	641642	1900-01-16	MA
Haines, James H.	Portable Basin	590833	1897-09-28	FL
Hammonds, Julia Terry	Apparatus for Holding Yarn Skeins	572985	1896-12-15	IL
Harding, Felix H.	Extension Banquet Table	614468	1898-11-22	OH
Harney, Michael C.	Lantern or Lamp	303844	1884-08-19	MO
Hawkins, Joseph	Gridiron	3973	1845-03-26	NJ
Hawkins, Randall	Harness Attachment	370943	1887-10-04	OH
Headen, M.	Foot Power Hammer	350363	1886-10-05	VA
Hearns, Robert	Sealing Attachment for Bottles	598929	1898-02-15	MN
Hearns, Robert	Detachable Car Fender	628003	1899-07-04	MN
Hearns, William	Device for Removing and Inserting Taps and Plugs in Water Mains	1040538	1912-10-08	DC
Hilyer, Andrew F.	Water Evaporator Attachment Hot Air Registers	435095	1890-08-26	DC
Hilyer, Andrew F.	Registers	438159	1890-10-14	DC
Hodges, Clarence and McCoy, Elijah J.	Steam Dome	320354	1885-06-16	MI
Holmes, Elijah H.	Gage	549513	1895-11-12	TX
Holmes, Lydia M.	The Knockdown Weeled Toy	2529692	1950-11-14	FL
Huges, Isaiah D.	Combined Excavator and Elevator	687312	1901-11-26	KS
Hunter, John W.	Portable Weighing Scales	570553	1896-11-03	IA

Hyde, Robert N.	Composition for Cleaning & Preserving Carpets	392205	1883-11-06	IA
Jackson, Benjamin F.	Heating Apparatus	599985	1898-03-01	MA
Jackson, Benjamin F.	Matrix Drying Apparatus	603879	1898-05-10	MA
Jackson, Benjamin F.	Gas Burner	622482	1899-04-04	MA
Jackson, Benjamin F.	Steam Boiler	690730	1902-01-07	MA
Jackson, Henry A.	Kitchen Table	569135	1896-10-06	DC
Jackson, William H.	Railway Switch	578641	1897-03-09	IN
Jackson, William H.	Railway Switch	593665	1897-03-16	IN
Jackson, William H.	Automatic Locking Switch	609436	1898-08-23	IN
Jackson, William H.	Overcoming Dead Centers	612345	1898-10-11	IN
Jennings, Thomas L.	Dry scouring	0	1821-03-03	NY
Johnson, Daniel	Rotary Dining Table	396089	1889-01-15	MO
Johnson, Daniel	Lawn Mower Attachment	410836	1889-09-10	MO
Johnson, Daniel	Grass Receivers for Lawn Mowers	429629	1890-06-10	MO
Johnson, Issac R.	Bicycle Frame	634823	1899-10-10	NY
Johnson, Payton	Swinging Chairs	249530	1881-11-15	OH
Johnson, Powell	Eye Protector	234039	1880-11-02	AL
Johnson, Willie H.	Overcoming Dead Centers	554223	1896-02-04	TX
Johnson, Willis	Egg Beater	292821	1884-02-05	OH
Johnson, Willis	Paint Vehicle	393763	1888-12-04	OH
Johnson, Willis	Velocipede	627335	1899-06-20	OH
Jones, Frederick M.	Ticket Dispensing Machine	2163754	1939-06-27	MN
Jones, Frederick M.	Means for Automatically Stopping & Starting Gas Engines	2337164	1943-12-21	MN
Jones, Frederick M.	Two-Cycle Gas Engine	2376968	1945-05-29	MN
Jones, Frederick M.	Two-Cycle Gas Engine	2417253	1947-03-11	MN
Jones, Frederick M.	Air Conditioning Unit	2475841	1949-07-12	MN
Jones, Frederick M.	Starter Generator	2475842	1949-07-12	MN
Jones, Frederick M.	Means for Thermostatically Operating Gas Engines	2477377	1949-07-26	MN
Jones, Frederick M.	Rotary Compressor	2504841	1950-04-18	MN
Jones, Frederick M.	System for Controlling Operation of Refrigeration	2509099	1950-05-23	MN

	Units			
Jones, Frederick M.	Apparatus for Heating or Cooling Atmosphere in Enclosures	2526874	1950-10-24	MN
Jones, Frederick M.	Two-Cycle Gasoline Engine	2523273	1950-11-28	MN
Jones, Frederick M.	Prefabricated Refrigerator Construction	2535682	1950-12-26	MN
Jones, Frederick M.	Refrigeration Control Device	2581956	1952-01-08	MN
Jones, Frederick M.	Methods and Means of Defrosting a Cold Diffuser	2666298	1954-01-19	MN
Jones, Frederick M.	Method for Air Conditioning	2696086	1954-12-07	MN
Jones, Frederick M.	Method for Preserving Perishables	2780923	1957-02-12	MN
Jones, Frederick M.	Control Device for Internal Combustion Engine	2850001	1958-09-02	MN
Jones, Frederick M.	Thermostat and Temperature Control System	2926005	1960-02-23	MN
Jones, James C.	Portable Drill Frame	532881	1895-01-22	PA
Joyce, James A.	Ore Bucket	603143	1898-04-26	OH
Kelley, George W.	Steam Table	592591	1897-10-26	VA
Latimer, Lewis & Nichols, Joseph H	Electric Lamp	247097	1881-09-13	NY
Latimer, Lewis H.	Manufacturing Carbons	252386	1882-01-17	NY
Latimer, Lewis H.	Apparatus for Cooling and Disinfecting	334078	1886-01-12	NY
Latimer, Lewis H.	Locking Racks for Coats, Hats, and Umbrellas	557076	1896-03-24	NY
Latimer, Lewis H.	Book Supporter	781890	1905-02-07	NY
Lavalette, William A.	Printing Press	208184	1878-09-17	DC
Lee, Henry	Animal Trap	61941	1867-02-12	OH
Lee, Joseph	Kneading Machine	524042	1894-08-07	MA
Lee, Joseph	Bread Crumbing Machine	540553	1895-06-04	MA
Leslie, Frank W.	Envelope Seal	590325	1897-09-21	IL
Lewis, Anthony L.	Window Cleaner	483359	1892-09-27	IL
Lewis, Edward R.	Spring Gun	362096	1887-05-03	MA
Linden, Henry	Piano Truck	459365	1891-09-08	OH
Little, Ellis	Bridle-Bit	254666	1882-03-07	NY
Long, Amos E. and Jones, Albert A.	Caps for Bottles	610715	1898-09-13	PA

Loudin, Frederick J.	Sash Fastener	510432	1893-12-12	OH
Loudin, Frederick J.	Key Fastener	512308	1894-01-09	OH
Love, John L.	Plasterers' Hawk	542419	1895-07-09	MA
Love, John L.	Pencil Sharpener	594114	1897-11-23	MA
Madison, Walter G.	Flying Machine	1047098	1912-12-10	IA
Marshall, Willis	Grain Binder	341589	1886-05-11	IL
Marshman, Newman R. and Burridge, L. S.	Type Writing Machine	315386	1885-04-07	NY
Martin, Thomas J.	Fire Extinguisher	125063	1872-05-26	MI
Martin, Washington A.	Lock	407738	1889-07-23	IL
Martin, Washington A.	Lock	443945	1890-12-30	IL
Matzeliger, Jan E.	Shoe Lasting Machine	274207	1883-03-20	MA
Matzeliger, Jan E.	Mechanism for Distributing Tacks	415726	1889-11-26	MA
Matzeliger, Jan E.	Nailing Machine	421954	1890-02-25	MA
Matzeliger, Jan E.	Tack Separating Mechanism	423937	1890-03-25	MA
Matzeliger, Jan E.	Lasting Machine	459899	1891-09-22	MA
McCoy & Hodges	Lubricator	418139	1889-12-24	MI
McCoy, Elijah J.	Lubricator for Steam Engines	129843	1872-07-02	MI
McCoy, Elijah J.	Lubricator for Steam Engines	130305	1872-08-06	MI
McCoy, Elijah J.	Lubricator	139407	1873-05-27	MI
McCoy, Elijah J.	Steam Lubricator	146697	1874-01-20	MI
McCoy, Elijah J.	Ironing Table	150876	1874-05-12	MI
McCoy, Elijah J.	Steam Cylinder Lubricator	173032	1876-02-01	MI
McCoy, Elijah J.	Steam Cylinder Lubricator	179585	1876-07-04	MI
McCoy, Elijah J.	Lubricator	255443	1882-03-28	MI
McCoy, Elijah J.	Lubricator	261166	1882-07-18	MI
McCoy, Elijah J.	Lubricator	320379	1885-06-16	MI
McCoy, Elijah J.	Lubricator	357491	1887-02-08	MI
McCoy, Elijah J.	Lubricator Attachment	361435	1887-04-19	MI
McCoy, Elijah J.	Lubricator for Safety Valves	363529	1887-05-24	MI
McCoy, Elijah J.	Lubricator	383745	1888-05-29	MI
McCoy, Elijah J.	Lubricator	383746	1888-05-29	MI
McCoy, Elijah J.	Drip Cup	460215	1891-09-29	MI

McCoy, Elijah J.	Lubricator	465875	1891-12-29	MI
McCoy, Elijah J.	Lubricator	472066	1892-04-05	MI
McCoy, Elijah J.	Lubricator	610634	1898-09-13	MI
McCoy, Elijah J.	Lubricator	611759	1898-10-04	MI
McCoy, Elijah J.	Oil Cup	614307	1898-11-15	MI
McCoy, Elijah J.	Lubricator	627623	1899-06-27	MI
McCree, Daniel	Portable Fire Escape	440322	1890-11-11	IL
McNair, Luther	Sanitary Attachment for Drinking Glasses	1034636	1912-08-06	NY
Mendenhall, Albert	Holder for Driving Reins	637811	1899-11-28	OH
Miles, Alexander	Elevator	371207	1887-10-11	MN
Mitchell, Charles L.	Phneterisin	291071	1884-01-01	MA
Mitchell, James M.	Cheek Row Corn Planter	641462	1900-01-16	KY
Morehead, King	Reel Carrier	568916	1896-10-06	MO
Morgan, Garrett A.	Breathing Device	1113675	1914-10-13	OH
Morgan, Garrett A.	Traffic Signal	1475024	1923-11-20	OH
Murray, George W.	Combined Furrow Opener and Stalk-Knocker	517960	1894-04-10	SC
Murray, George W.	Cultivator and Marker	517961	1894-04-10	SC
Murray, George W.	Planter	520887	1894-06-05	SC
Murray, George W.	Cotton Chopper	520888	1894-06-05	SC
Murray, George W.	Fertilizer Distributor	520889	1894-06-05	SC
Murray, George W.	Planter	520890	1894-06-05	SC
Murray, George W.	Combined Cotton Seed	520891	1894-06-05	SC
Murray, George W.	Planter and Fertilizer Distributor Reaper	520892	1894-06-05	SC
Murray, William	Corn Harvester	99463	1870-02-01	VA
Murray, William	Attachment for Bicycles	445452	1891-01-27	VA
Nance, Lee	Game Apparatus	464035	1891-12-01	NJ
Nash, Henry H.	Life Preserving Stool	168519	1875-10-05	MD
Newman, Lyda D.	Improvement on the Hair Brush	614335	1898-11-15	NY
Newson, S.	Oil Heater or Cooker	520188	1894-05-22	SC
Nickerson, William J.	Mandolin and Guitar Attachment for Pianos	627739	1899-06-27	LA
O'Connor & Turner	Alarm for Boilers	566612	1896-08-25	NY

O'Connor & Turner	Steam Gage	566613	1896-08-25	NY
O'Connor & Turner	Alarm for Coasts Containing Vessels	598572	1898-02-08	NY
Outlaw, John W.	Horseshoes	614273	1898-11-15	NY
Parker, Alice H.	Improvement on the Heating Furnace	1325905	1919-12-23	NJ
Parker, John P.	Follower Screw for Tobacco Presses	304552	1884-09-02	OH
Parker, John P.	Portable Screw Press	318285	1885-05-19	OH
Payne, Moses	Horseshoe	394388	1888-12-11	KY
Pelham, Robert A.	Pasting Apparatus	807685	1905-12-19	MI
Perryman, Frank R.	Caterers' Tray Table	468038	1892-02-02	IL
Peterson, Henry	Attachment for Lawn Mowers	402189	1889-04-30	CA
Phelps, William H.	Apparatus for Washing Vehicles	579242	1897-03-23	IN
Pickering, John F.	Air Ship	643975	1900-02-20	H
Pickett, Henry	Scaffold	152511	1874-06-30	OH
Pinn, Traverse. B.	File Holder	231355	1880-08-17	VA
Polk, Austin J.	Bicycle Support	558103	1896-04-14	IL
Pugsley, Abraham	Blind Stop	433306	1890-07-29	RI
Pugsley, Abraham	Blind Stop	433819	1890-08-05	RI
Pugsley, Samuel	Gate Latch	357787	1887-02-15	NY
Purdy & Sadgwar	Folding Chair	405117	1889-06-11	DC
Purdy, Walter	Device for Sharpening Edged Tools	570337	1896-10-27	PA
Purdy, Walter	Design for Sharpening Edged Tools	609367	1898-08-16	PA
Purdy, Walter	Device for Sharpening Edged Tools	630106	1899-08-01	PA
Purvis, William B.	Bag Fastener	256856	1882-04-25	PA
Purvis, William B.	Hand Stamp	273149	1883-02-27	PA
Purvis, William B.	Paper Bag Machine	293353	1884-02-12	PA
Purvis, William B.	Fountain Pen	419065	1890-01-07	PA
Purvis, William B.	Paper Bag Machine	420099	1890-01-28	PA
Purvis, William B.	Paper Bag Machine	430684	1890-06-24	PA
Purvis, William B.	Paper Bag Machine	434461	1890-08-19	PA
Purvis, William B.	Paper Bag Machine	435524	1890-09-02	PA

Purvis, William B.	Paper Bag Machine	460093	1891-09-22	PA
Purvis, William B.	Electric Railway	519291	1894-05-01	PA
Purvis, William B.	Paper Bag Machine	519348	1894-05-08	PA
Purvis, William B.	Paper Bag Machine	519349	1894-05-08	PA
Purvis, William B.	Magnetic Car Balancing Device	539542	1895-05-21	PA
Purvis, William B.	Paper Bag Machine	578361	1897-03-09	PA
Purvis, William B.	Electric Railway Switch	588176	1897-08-17	PA
Queen, William	Guard for Companion Ways and Hatches	458131	1891-08-18	MD
Ray, Lloyd P.	Dust Pan	587607	1897-08-03	WA
Ray, Lloyd P.	Chair Supporting Device	620078	1899-02-21	WA
Reed, Judy W.	Improvements in the Doug Kneader and Roller	305474	1884-09-20	DC
Reynolds, Humphrey H.	Window Ventilator for Railroad Cars	275271	1883-04-03	MI
Reynolds, Humphrey H.	Safety Gate for Bridges	437937	1890-10-07	MI
Reynolds, Mary Jane	Hoisting and Loading Mechanism	1337667	1920-04-20	MO
Reynolds, Robert R.	Non-Refillable Bottle	624092	1899-05-02	NY
Rhodes, Jerome B.	The Water Closet	639290	1899-12-19	LA
Richardson, Albert C.	Hame Fastener	255022	1882-03-14	MI
Richardson, Albert C.	Churn	446470	1891-02-17	MI
Richardson, Albert C.	Casket Lowering Device	529311	1894-11-13	MI
Richardson, Albert C.	Insect Destroyer	620362	1899-02-28	MI
Richardson, Albert C.	Bottle	638811	1899-12-12	MI
Richardson, William H.	Cotton Chopper	343140	1886-06-01	MD
Richardson, William. H.	Child's Carriage	405599	1889-06-18	MD
Richardson, William. H.	Child's Carriage	405600	1889-06-18	MD
Richey, Charles V.	Car Coupling	584650	1897-06-15	DC
Richey, Charles V.	Railroad Switch	587657	1897-08-03	DC
Richey, Charles V.	Railroad Switch	592448	1897-10-26	DC
Richey, Charles V.	Fire Escape Bracket	596427	1897-12-28	DC
Richey, Charles V.	Combined Hammock and Stretcher	615907	1898-12-13	DC
Richey, Charles V.	Telephone Register and Lock Out Device	1063599	1913-06-03	DC

Rickman, Alvin L.	Overshoe	598816	1898-02-08	OH
Ricks, James	Horseshoe	338781	1886-03-30	DC
Ricks, James	Overshoes for Horses	626245	1899-06-06	DC
Rillieux, Norbert	Sugar Refiner (Vacuum Pan)	3237	1843-08-26	LA
Rillieux, Norbert	Sugar Refiner (Evaporating Pan)	4879	1846-12-10	LA
Robinson, E. R.	Electric Railway Trolley	505370	1893-09-19	TN
Robinson, Elbert R.	Casting Composite	594286	1897-11-23	TN
Robinson, James H.	Life Saving Guards for Locomotives	621143	1899-03-14	MN
Robinson, James H.	Life Saving Guards for Street Cars	623929	1899-04-25	MN
Robinson, John	Dinner Pail	356852	1887-02-01	WV
Romain, Arnald	Passenger Register	402035	1889-04-23	LA
Ross, A. L.	Runner for Stops	565301	1896-08-04	NY
Ross, A. L.	Bag Closure	605343	1898-06-07	NY
Ross, A. L.	Trousers Support	638068	1899-11-28	NY
Ross, Joseph	Bailing Press	632539	1899-09-05	IN
Rosten, David N.	Feather Curler	556166	1896-03-10	CT
Russell, Lewis A.	Guard Attachment for Beds	544381	1895-08-13	MO
Sammons, Walter H.	The Comb	1362823	1920-12-21	PA
Sampson, George T.	Sled Propeller	312388	1885-02-17	OH
Sampson, George T.	Clothes Drier	476416	1892-06-07	OH
Scott, Robert P.	Corn Silker	524223	1894-08-07	OH
Scottron, Samuel R.	Adjustable Window Cornice	224732	1880-02-17	NY
Scottron, Samuel R.	Cornice	270851	1883-01-16	NY
Scottron, Samuel R.	Pole Tip	349525	1886-09-30	NY
Scottron, Samuel R.	Curtain Rod	481720	1892-08-30	NY
Scottron, Samuel R.	Supporting Bracket	505008	1893-09-12	NY
Shanks, Stephen C.	Sleeping Car Berth Register	587165	1897-07-21	C
Shorter, Dennis W.	Feed Rack	363089	1887-05-17	NY
Smartt, Brinay	Reversing Valve	799498	1905-09-12	TN
Smartt, Brinay	Valve Gear	935169	1909-09-28	TN
Smartt, Brinay	Wheel	1052290	1913-02-04	TN

Smith, John W.	Improvement in Games	647887	1900-04-17	MA
Smith, Joseph H.	Lawn Sprinkler	581785	1897-05-04	MA
Smith, Joseph H.	Lawn Sprinkler	601065	1898-03-22	MA
Smith, Mildred E.	Family Relationships Card Game	4230321	1980-10-28	DC
Smith, Peter D.	Potato Digger	445206	1891-01-21	OH
Smith, Peter D.	Grain Binder	469279	1892-02-23	OH
Snow, William and Johns, James A.	Liniment	437728	1890-10-07	DC
Spears, Harde	Portable Shield for Infantry	110599	1870-12-27	NC
Spikes, Richard B.	Billiard Cue Rack	972277	1910-10-11	NM
Spikes, Richard B.	Combination Milk Bottle Opener & Bottle Cover	1590557	1926-06-29	NM
Spikes, Richard B.	Apparatus for Obtaining Samples & Temperature of Liquids	1828753	1931-10-27	CA
Spikes, Richard B.	Automatic Gear Shift	1889814	1932-12-06	CA
Spikes, Richard B.	Transmission and Shifting Thereof	1936996	1933-11-28	CA
Standard, John	Oil Stove	413689	1889-10-29	NJ
Standard, John	Refrigerator	455891	1891-07-14	NJ
Stewart, Enos W.	Punching Machine	362190	1887-05-03	MI
Stewart, Enos W.	Machine for Forming Vehicle Sear Bars	373698	1887-11-22	MI
Stewart, Thomas W.	Metal Bending Machine	375512	1887-12-27	MI
Stewart, Thomas W.	Mop	499402	1893-06-13	MI
Stewart, Thomas W.	Station Indicator	499895	1893-06-20	MI
Sutton, Edward H.	Cotton Cultivator	149543	1874-04-07	NC
Sweeting, James A.	Device for Rolling Cigarettes	594501	1897-11-30	NY
Sweeting, James A.	Combined Knife and Scoop	605209	1898-06-07	NY
Taylor, Benjamin H.	Rotary Engine	202888	1878-04-23	MS
Taylor, Benjamin H.	Slide Valve	585798	1897-07-06	MS
Thomas, Samuel E.	Waste Trap	286746	1883-10-16	NY
Thomas, Samuel E.	Waste Trap for Basins, Closets, Etc.	371107	1887-10-04	NY
Thomas, Samuel E.	Casting	386941	1888-07-31	NY
Thomas, Samuel E.	Pipe Connection	390821	1888-10-09	NY

Toliver, George	Propeller for Vessels	451086	1891-04-28	PA
Tregoning & Latimer	Globe Supporter for Electric Lamps	255212	1882-03-21	NY
Turner, M. Turner	The Fruit Press	1180959	1916-04-25	CA
Virgle M. Ammons	Fireplace Damper Actuating Tool	3908633	1975-09-30	WV
Walker, Peter	Machine for Cleaning Seed Cotton	577153	1897-02-16	MS
Walker, Peter	Bait Holder	600241	1898-03-08	MS
Waller, Joseph W.	Shoemaker's Cabinet or Bench	224253	1880-02-03	MD
Washington, W.	Corn Husking Machine	283173	1883-08-14	WV
Watkins, Isaac	Scrubbing Frame	437849	1890-10-07	PA
Watts, Julius R.	Bracket for Miners' Lamp	493137	1893-03-07	IL
Webb, Henry C.	Saw Attachment	483971	1892-10-04	ID
West, Edward H.	Weather Shield	632385	1899-09-05	MA
West, John W.	Running Gear (Wagon)	108419	1870-10-18	IA
White, Daniel L.	Extension Steps for Cars	574969	1897-01-12	OH
White, John T.	Lemon Squeezer	572849	1896-12-08	NY
Williams, C.	Canopy Frame	468280	1892-02-02	PA
Williams, James P.	Pillow Sham Holder	634784	1899-10-10	MA
Winn, Frank	Direct Acting Steam Engine	394047	1888-12-04	TX
Winters, Joseph R.	Fire Escape Ladders	203517	1878-05-07	PA
Winters, Joseph R.	Fire Escape Ladder	214224	1879-04-08	PA
Wood, Francis J.	Potato Digger	537953	1895-04-23	MI
Woods, Granville T.	Steam Boiler Furnace	299894	1884-06-03	OH
Woods, Granville T.	Telephone Transmitter	308817	1884-12-02	OH
Woods, Granville T.	Apparatus for Transmission of Messages by Electricity	315368	1885-04-07	OH
Woods, Granville T.	Relay Instrument	364619	1887-06-07	OH
Woods, Granville T.	Polarized Relay	366192	1887-07-05	OH
Woods, Granville T.	Electromechanical Brake	368265	1887-08-16	OH
Woods, Granville T.	Telephone System and Apparatus	371241	1887-10-11	OH
Woods, Granville T.	Electromagnetic Brake Apparatus	371655	1887-10-18	OH
Woods, Granville T.	Railway Telegraphy	373383	1887-11-15	OH

Woods, Granville T.	Induction Telegraph System	373915	1887-11-29	OH
Woods, Granville T.	Overhead Conducting System for Electric Railway	383844	1888-05-29	OH
Woods, Granville T.	Electromotive Railway System	385034	1888-06-26	OH
Woods, Granville T.	Tunnel Construction for Electric Railway	386282	1888-07-17	OH
Woods, Granville T.	Galvanic Battery	387839	1888-08-14	OH
Woods, Granville T.	Railway Telegraphy	388803	1888-08-28	OH
Woods, Granville T.	Automatic Cut-off Switch	395533	1889-01-01	OH
Woods, Granville T.	Automatic Cut-off Switch	438590	1890-10-14	OH
Woods, Granville T.	Electric Railway System	463020	1891-11-10	OH
Woods, Granville T.	Electric Railway Supply System	507606	1893-10-31	OH
Woods, Granville T.	Electric Railway Conduit	509065	1893-11-21	OH
Woods, Granville T.	System of Electrical Distribution	569443	1896-10-13	OH
Woods, Granville T.	Amusement Apparatus	639692	1899-12-19	OH
Woods, Granville T.	Incubator	656760	1900-08-28	OH
Woods, Granville T.	Electric Railway	662049	1900-11-20	OH
Woods, Granville T.	Electric Railway	667110	1901-01-29	OH
Woods, Granville T.	Electric Railway System	678086	1901-07-09	OH
Woods, Granville T.	Regulating and Controlling Electrical Translating Devices	681768	1901-09-03	OH
Woods, Granville T.	Electric Railway	687098	1901-11-19	OH
Woods, Granville T.	Apparatus for Controlling Electric Motors	690808	1902-01-07	OH
Woods, Granville T.	Electric Railway	695988	1902-03-25	OH
Woods, Granville T.	Automatic Air Brake	701981	1902-06-10	OH
Woods, Granville T.	Electric Railway System	718183	1903-01-13	OH
Woods, Granville T.	Electric Railway	729481	1903-05-26	OH
Woods, Granville T.	Electric Railway Apparatus	762792	1904-06-14	OH
Wormley, James	Life Saving Apparatus	242091	1881-05-24	DC

BIBLIOGRAPHY

Dodo Press, editor. “The Colored Inventor: A Record of Fifty Years, and the Negro in the Field of Invention.” The Colored Inventor, Dodo Press, 1913, pp. 1–2. Dodo Press, www.dodopress.co.uk.

American Experience, PBS. “Summing up, Looking Forward and the Paris Exposition.” American Experience | PBS, 29 Mar. 2019, www.pbs.org/wgbh/americanexperience/features/1900-forward-exposition.

“Paris 1900 Exposition: History, Images, Interpretation — Ideas.” Ideas, www.arthurchandler.com/paris-1900-exposition. Accessed 7 Apr. 2025.

Cmires. “Pennsylvania Emancipation Exposition (1913).” Encyclopedia of Greater Philadelphia, 9 Dec. 2024, philadelphiaencyclopedia.org/essays/pennsylvania-emancipation-exposition-1913. Accessed 7 Apr. 2025.

“Found on Baker’s List.” USPTO, 14 Jan. 2025, www.uspto.gov/learning-and-resources/journeys-innovation/historical-stories/found-bakers-list. Accessed 13 Mar. 2025.

Wikipedia contributors. “Exposition Universelle (1900).” Wikipedia, 8 Mar. 2025, en.wikipedia.org/wiki/Exposition_Universelle_(1900). Accessed 7 Apr. 2025.

“Henry Blair (Inventor).” Wikipedia, 17 Mar. 2025, en.wikipedia.org/wiki/Henry_Blair_(inventor). Accessed 30 Mar. 2025.

“Henry E. Baker.” Wikipedia, 30 Oct. 2024, en.wikipedia.org/wiki/Henry_E._Baker. Accessed 11 Mar. 2025.

“Crisis : Du Bois, W. E. B. (William Edward Burghardt), 1868-1963, Ed : Free Download, Borrow, and Streaming : Internet Archive.” Internet Archive, 1910, archive.org/details/crisis0506dubo/page/158/mode/2up?q=Exposition. Accessed 8 Apr. 2025.

“African American Photographs Assembled for 1900 Paris Exposition - Du Bois Materials - Prints and Photographs Online Catalog (Library of Congress).” Library of Congress, www.loc.gov/pictures/collection/anedub/dubois.html. Accessed 8 Apr. 2025.

Patent Public Search Basic | USPTO. ppubs.uspto.gov/pubwebapp/static/pages/ppubsbasic.html. Accessed 11 Mar. 2025.

Wikipedia contributors. “1836 U.S. Patent Office Fire.” Wikipedia, 23 Jan. 2025, en.wikipedia.org/wiki/1836_U.S._Patent_Office_fire. Accessed 8 Apr. 2025.

Licensing the First US Patent | Lemelson. invention.si.edu/invention-stories/licensing-first-us-patent#:~:text=On%20July%2031%2C%201790%2C%20the,as%20an%20ingredient%20in%20soap. Accessed 14 Apr. 2025.

“History of United States Patent Law.” Wikipedia, 15 Sept. 2024, en.wikipedia.org/wiki/History_of_United_States_patent_law. Accessed 14 Apr. 2025.

AROUT THE AUTHOR

Reggie K. Smith

Reggie K. Smith is a creative technology innovator—a seasoned expert in video production technology and graphic communications, with four decades of experience in the tech and creative industries. A native of Mississippi and a lifelong advocate for innovation, Reggie has spent the last 20 years studying and documenting the contributions of African American inventors. Now, in a bold career pivot, Reggie brings this wealth of research to the literary world with the second of three publications, *BAKER'S LIST Of "Colored" Inventors: Over 400 Innovations by African Americans (1821–1980)*—a compelling collection that illuminates the ingenuity and resilience of and places within your grasp hundreds of Black inventors across generations.

The third work delves deeper as it explores Henry E. Baker's life and journey, complete with biographical sketches of inventors, educational patent detail, and more, in a way that nicely blends creativity and technology for an engaging experience. You will be empowered along your own educational journey.

Made in the USA
Coppell, TX
23 January 2026